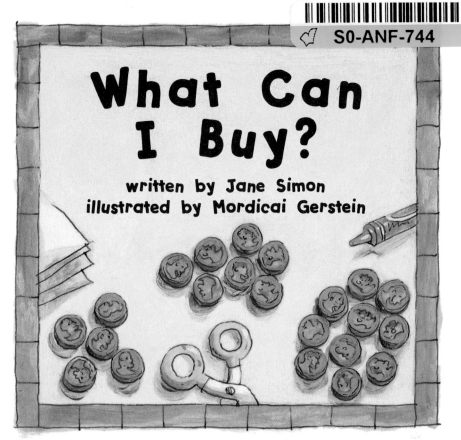

What Can I Buy?

written by Jane Simon
illustrated by Mordicai Gerstein

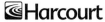

Orlando Boston Dallas Chicago San Diego

www.harcourtschool.com

S0-ANF-744

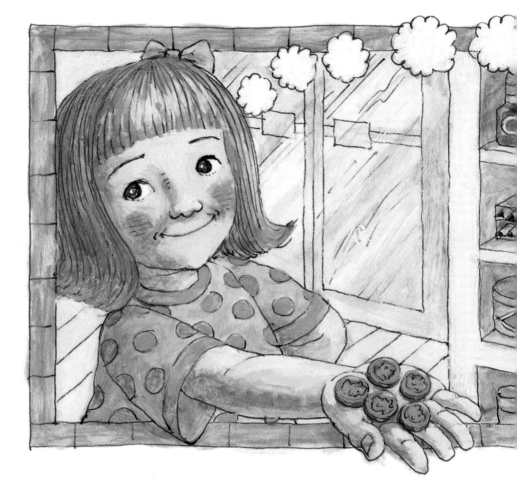

5 pennies

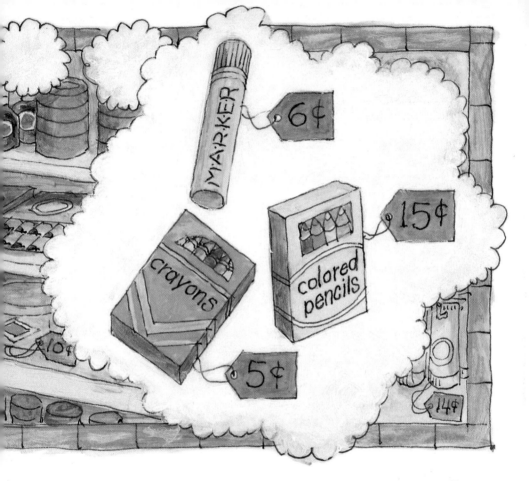

What can I buy?

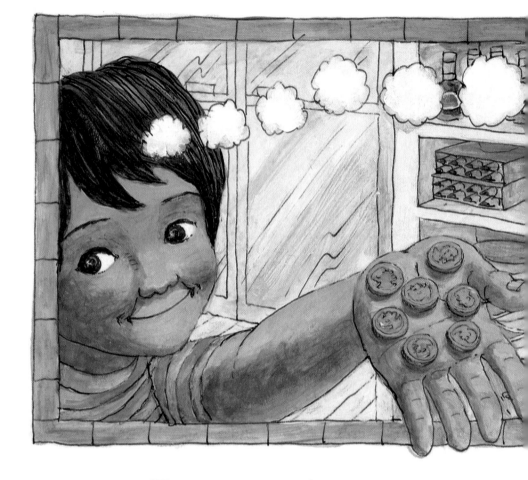

7 pennies

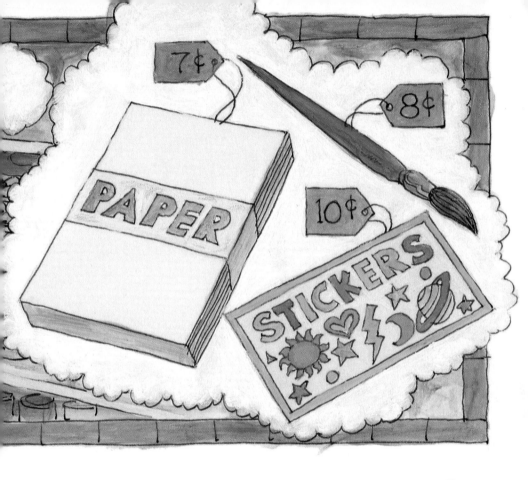

What can I buy?

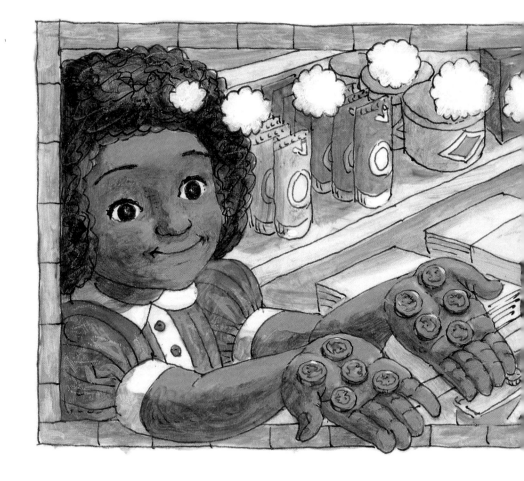

10 pennies

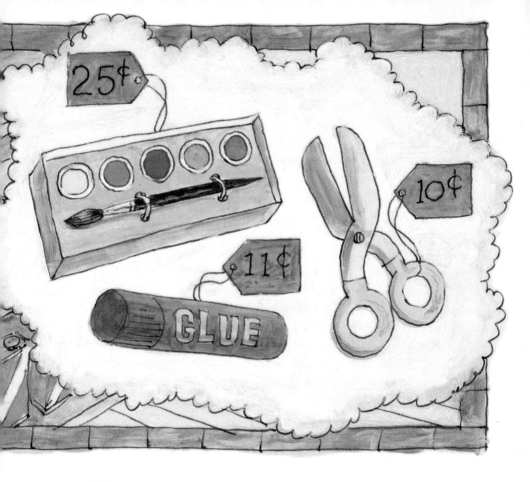

What can I buy?

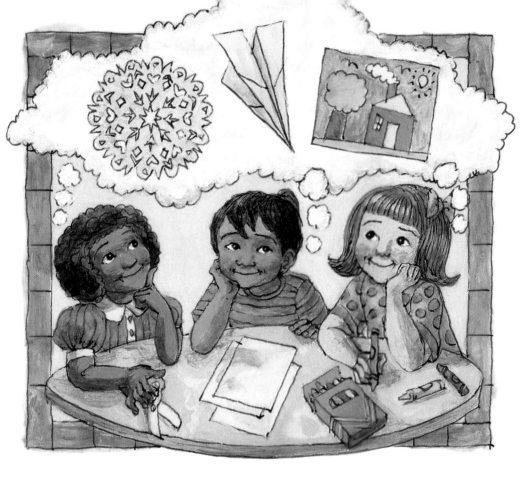

What can I make?